ISTWA NIMEWO

THE NUMBER STORY

SMALL BOOK ONE

ENGLISH - HAITIAN CREOLE

*Numbers Teach Children
Their Number Names*

written and illustrated by

MISS ANNA

Early Reader Edition of *The Number Story 1*
Bronze Medal Winner, 2016 Wishing Shelf Book Award

Library of Congress Control Number: 2018902040

Names: Miss Anna, author.
Title: Number story : numbers teach children their number names / Miss Anna.
Description: Portland, OR: Lumpy Publishing, 2018.
Identifiers: ISBN 978-1-949320-15-2| LCCN 2018902040
Summary: The pictures and rhymes present stories which introduce numbers 0-10.
Subjects: LCSH Numeration—English—Haitian Creole--Pictorial works--Juvenile literature. |
BISAC JUVENILE NONFICTION /
Languages: English—Haitian Creole
Classification: LCC QA141.3 .M57 2018 | DDC 513—dc23

Publisher: Lumpy Publishing
Website: www.missannabooks.com
Email: missanna@missannabooks.com

Paperback: ISBN 978-1-949320-15-2
Printed in the U.S.A.   1 3 5 7 9 10 8 6 4 2

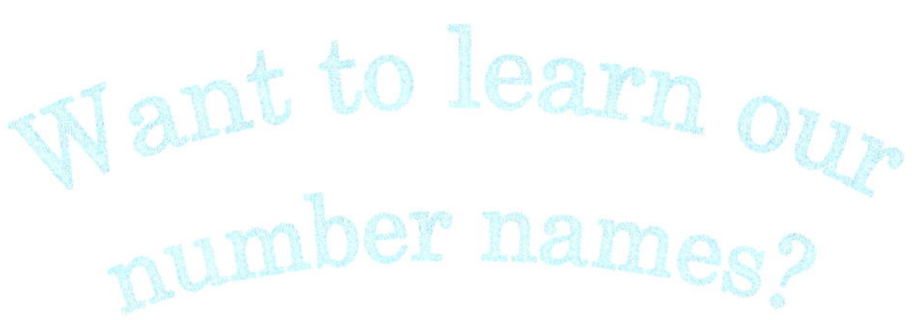

Èske ou ta renmen aprann
non nimewo yo?

It is very easy and a lot of fun!

Li trè fasil ak li bay anpil plezi!

Say-along our little jingle

Chante avèk nou ti istwa nou!

starting from Number One!

Nou pral kòmanse nan Nimewo Youn!

1

ONE looks like my one finger.

YOUN sanble ak yon sèl dwèt mwen.

ONE!
YOUN!

2

TWO   trails a tail.

DE

lap trennen yon ke.

A TAIL! YON KE!

3

THREE has bumps.

TWA

gen anpil chòk.

BUMPY! AJITE!

4

FOUR carries a sail.

KAT

pote yon vwal.

4
A SAIL!
YON VWAL!

5

FIVE   is a racing track.

SENK

se yon pis pou kous.

VROOM
VOOOM!
1

6

SIX curves like a snail.

SIS

kwochi tankou yon kalmason.

YON KALMASON!

7

SEVEN has a sharp angle.

SET

li gen yon ang byen file.

BE CAREFUL! IT'S SHARP!
ATANSYON! LI FILE WI!

8

OURA! 
YIPIPIP!
YIPPEE!

9

NINE is a bubble on a stick.

NEF

se yon kim savon sou yon bwa.

A BUBBLE!

YOM KIM SAVON!

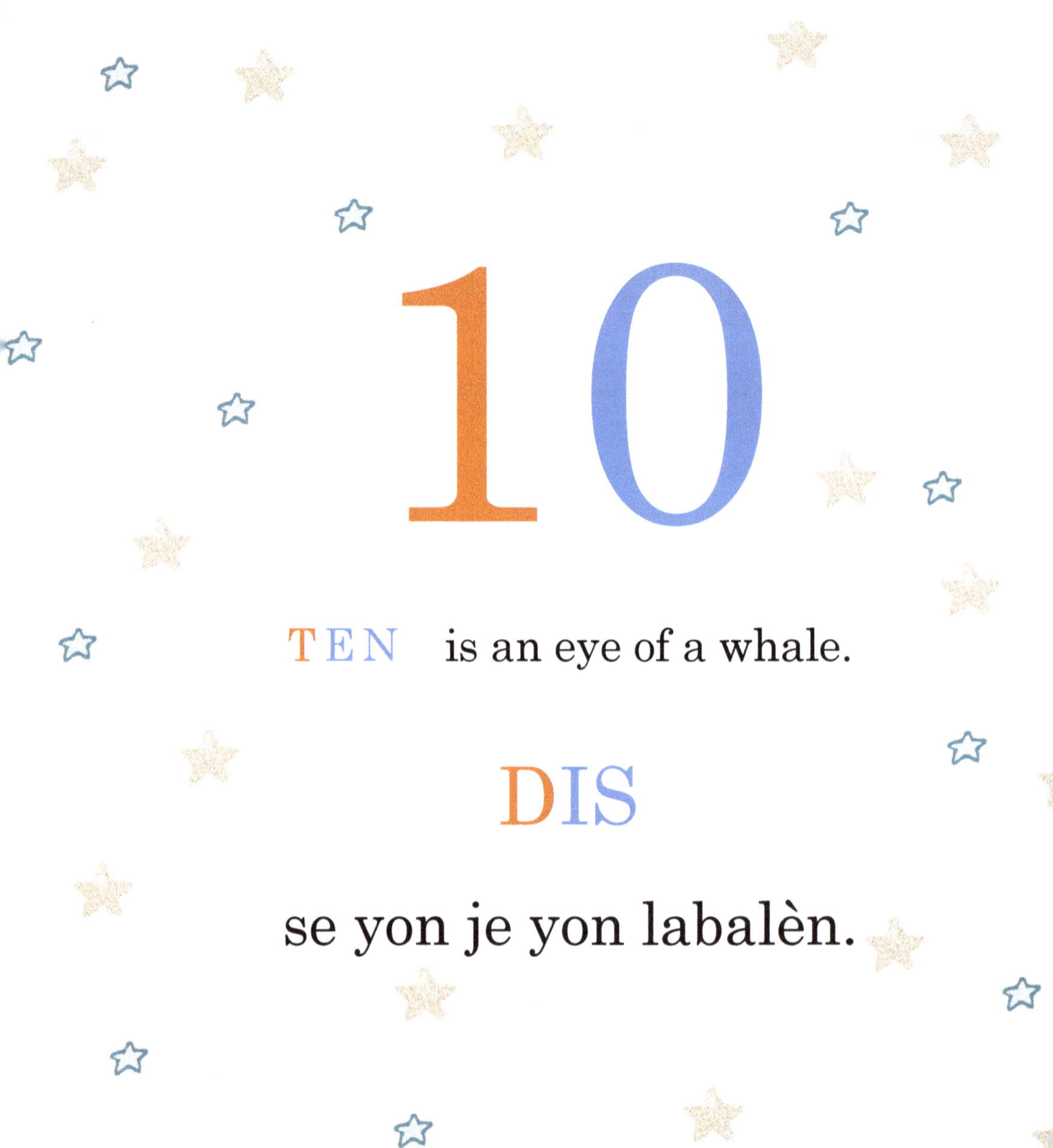
10
TEN is an eye of a whale.
DIS
se yon je yon labalèn.

HELLO!
ALLO!

And
Ak

O

ZERO   is an empty pail.

ZEWO

se yon bokit vid.

IT'S
EMPTY!
LIVID!

Thank you for playing with us today.

We had a lot of fun too!

Mèsi dèske wou te jwe avèk nou jodi a.

Nou te gen anpil plezi tou!

We are your Number friends,
Zero to Ten,
Who will be here for you~
Nou se zanmiw Nimewo yo
Zewo jiska Dis.
Nou ap toujou la pou wou.

Bye-bye now!
See you again soon!
Kounyeya Babay!
A byento nap wè ankò!

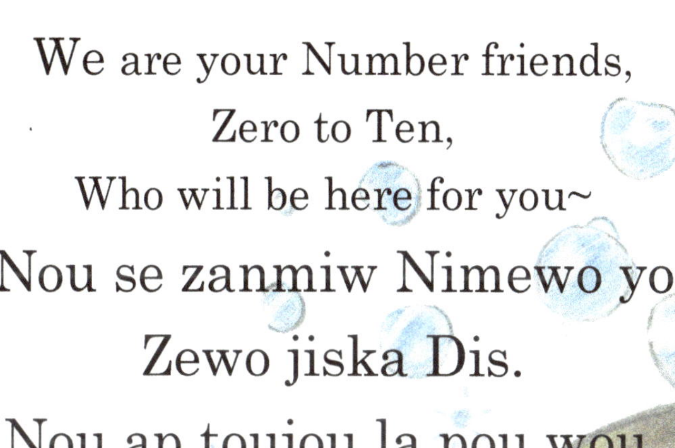

The Numbers are *SINGING* too!

To sing-a-long, look for Miss Anna Number Story
at your favorite music store like iTUNES.

MP3

Numbers 0-10
IDENTIFYING
& COUNTING

Numbers 11-20
& Ordinals

first, second, third...

Numbers 0-100
& Place Values

ones, tens, hundreds...

About Clock
& Telling Tim

hours, minutes, secon

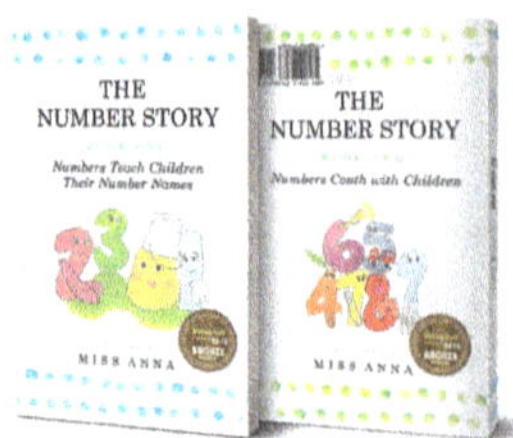

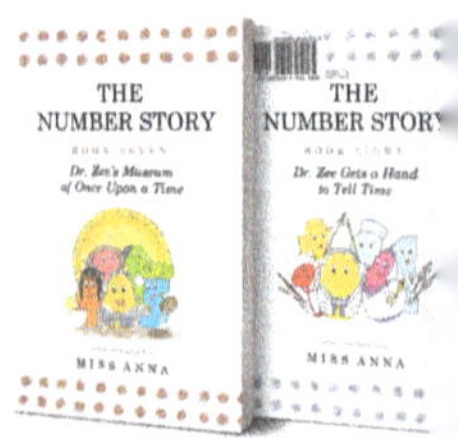

Number Story 1 & 2

isbn: 978-0-996216-48-7

Number Story 3 & 4

isbn: 978-1-945977-01-5

Number Story 5 & 6

isbn: 978-1-945977-06-0

Number Story 7 &

isbn: 978-1-949320-4

For more Miss Anna books to love,
visit us at

w w w . m i s s a n n a b o o k s . c o m

Numbers are working hard all over the world!
Come Travel the World with Us!

www.ingramcontent.com/pod-product-compliance
Lightning Source LLC
Chambersburg PA
CBHW041100050726
47599CB00018B/2215